EARLY BIRD
PHYSICS BOOKS

WHEELS
AND AXLES

by Sally M. Walker and Roseann Feldmann
photographs by Andy King

Lerner Publications Company • Minneapolis

For Sally Walker and Eileen Palsgrove, who made my dreams a reality—RF

The publisher wishes to thank the Minneapolis Kids program for its help in the preparation of this book.

Additional photographs are reproduced with permission from: © Dan Mahoney/Independent Picture Service, pp. 27, 46; © Jim McDonald/Corbis, p. 34; © Corbis Royalty Free, pp. 40, 42.

Text copyright © 2002 by Sally M. Walker and Roseann Feldmann
Photographs copyright © 2002 by Andy King, except where noted

Lerner Publications Company
A division of Lerner Publishing Group
241 First Avenue North
Minneapolis, MN 55401 U.S.A.

Website address: www.lernerbooks.com

Library of Congress Cataloging-in-Publication Data

Walker, Sally M.
  Wheels and axles / by Sally M. Walker and Roseann Feldmann ;
photographs by Andy King.
    p.   cm. — (Early bird physics books)
  Includes index.
  ISBN-13: 978-0-8225-2219-5 (lib. bdg. : alk. paper)
  ISBN-10: 0-8225-2219-5 (lib. bdg. : alk. paper)
  1. Wheels—Juvenile literature. 2. Axles—Juvenile literature.
[1. Wheels. 2. Axles] I. Feldmann, Roseann. II. King, Andy, ill. III. Title. IV. Series.
TJ181.5 W36 2002
621.8'11—dc21   00-011962

Manufactured in the United States of America
5 6 7 8 9 10 – JR – 10 09 08 07 06 05

CONTENTS

Be a Word Detective 5

Chapter 1 **WORK** 6

Chapter 2 **MACHINES** 10

Chapter 3 **FRICTION** 12

Chapter 4 **PARTS OF A WHEEL
AND AXLE** 22

Chapter 5 **GEARS** 36

A NOTE TO ADULTS
On Sharing a Book 44

LEARN MORE ABOUT
Simple Machines 45

Glossary 46

Index 47

BE A WORD DETECTIVE

Can you find these words as you read about wheels and axles? Be a detective and try to figure out what they mean. You can turn to the glossary on page 46 for help.

axle

complicated machines

force

friction

gear

simple machines

wheel

work

Playing soccer is work! What does the word "work" mean to a scientist?

Chapter 1

WORK

You work every day. At home, one of your chores may be painting. At school, you work when you sharpen your pencil.

You work at snack time. And you work when you race in gym. Eating and playing are work, too!

When scientists use the word "work," they don't mean the opposite of play. Work is using force to move an object. Force is a push or a pull. You use force to do chores, to play, and to eat.

You do work when you put newspapers in the recycling bin.

Anytime you use force to move an object to a new place, you have done work. Maybe the object moves hundreds of feet. Or maybe it only moves a tiny bit.

Riding a bike is work. Your force makes the bike's pedals turn. This makes you move forward.

You use force when you push your friend on a swing.

These kids are pushing very hard on a school building. But they are not doing work.

Pushing a school building is not work. It's not work even if you sweat. No matter how hard you try, you haven't done work. The building hasn't moved. If the building moves, then you have done work!

A train has many moving parts. What are machines that have many moving parts called?

Chapter 2
MACHINES

Most people want to make their work easy. Machines are tools that make work easier. Some machines make work go faster, too.

Some machines have many moving parts. They are complicated machines. Trains and cars are complicated machines.

Some machines have few moving parts. These machines are called simple machines. Simple machines are found in every home, school, and playground. These machines are so simple that most people don't realize they are machines.

 *The wheels on this chair are simple machines.*

*It is easy to make a
book slide on a table.
How can you make
this even easier?*

Chapter 3

FRICTION

Lay a book flat on a table. Give the book
a push. It's easy to make the book slide
a few inches. But you can make it even easier.

You'll need the book, a round pencil, a straw, a piece of paper, and a spool.

Put the round pencil under the book. Push the book again. A tiny push makes the book move easily. The pencil made your work easier. You used the pencil as a simple machine.

You'll need these objects to experiment with making work easier.

It is easier to push the book with a pencil under it.

It is easy to push the book when it lies flat on the table. But it is even easier to push the book with a pencil under it. This is because there is friction between the book and the tabletop. Friction is a force that makes a moving object slow down or stop. When the book is flat on the table, its whole side touches the table. The book slides only if your pushing force is stronger than friction's stopping force.

Look at the book while the pencil is under it. The pencil lifts the book above the table. Only one edge of the book touches the table. So there is almost no friction between the book and the table. That's why you don't have to push as hard. But there is still some friction. Where do you think there is still friction?

The pencil lifts most of the book above the table. So there is very little friction between the book and the table.

Take the book off the pencil. Look at the pencil. Part of the pencil touches the table. There's friction where the table and the pencil touch. You can prove it.

There is friction where the table and the pencil touch.

16

Friction makes the pencil stop rolling.

Put a piece of paper on the table. Put the pencil at one end of the paper. Push the pencil. What happens? It rolls. But then it stops. Friction makes the pencil stop. Try the same thing with the straw. The straw also rolls until friction stops it.

Draw a line in front of the straw.

Pretend there is no friction between the straw and the table. Then the straw would never stop rolling. What could you do to make less friction between the table and the straw?

Put your straw at one end of the paper. Draw a line right in front of the straw. Then roll the straw again. When the straw stops moving, pick it up. Put the pencil in its place. The pencil marks the straw's finish line.

Put the pencil in the straw's place. This way, you will mark the straw's finish line.

Slide the spool onto the straw.

You can lower the friction between the straw and the table. Slide the spool onto the straw. Now only the spool's narrow edges touch the table. Start from the same line you drew in front of the straw before. Push the spool and the straw. Try to use the same amount of force as you did before. What happens? The spool and the straw roll farther together than the straw rolled alone.

The spool and straw roll with enough force to push the pencil.

When the spool and the straw are together, they are a simple machine. What is this machine called?

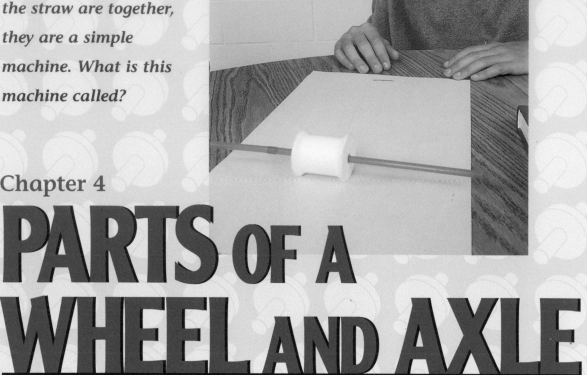

Chapter 4

PARTS OF A WHEEL AND AXLE

When you put a spool on a straw, you make a simple machine. It is called a wheel and axle. An axle goes through the center of a wheel. Your spool is a wheel. Your straw is an axle. The straw goes through the center of the spool.

22

Sometimes people use a wheel that spins around an axle. Hold your straw so it can't turn. Push the spool. The spool spins around the straw.

Try drawing a line around the spool. This will help you to see the spool spin.

You'll need a spool, a book, paper, a straw, and a thick pencil for the next experiment.

Sometimes people use a wheel and an axle that turn together. Take the spool off the straw. Put the spool on a thick pencil. It is all right if

24

the spool only fits on the sharpened end of the pencil. This time your pencil is the axle. The axle fits tightly into the wheel's center. You can't turn the wheel without turning the axle.

The spool should fit tightly on the pencil. If your pencil is too thin, wrap some masking tape around it to make it thicker.

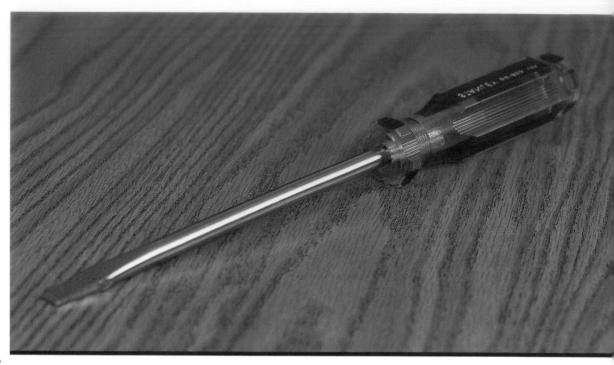

 A screwdriver is a wheel and axle. The handle is the wheel. The shaft is the axle.

Look at the wheel and axle made by the spool and the pencil. It looks like a screwdriver. A screwdriver is a wheel and axle. Its thick handle and thin metal shaft turn together to screw in a screw.

A doorknob is also a wheel and axle. The doorknob is the wheel. You can't see the axle.

26

The axle is inside the door. The doorknob and its axle turn together. When you turn the knob, it moves other parts inside the door. When those parts move, the door opens.

Faucet handles are wheels. When you turn the handles, other parts inside the faucet move. Then water flows.

Each time you turn a screwdriver or a doorknob, its axle turns. The outside of the handle travels a much longer distance than the axle does. You can prove this.

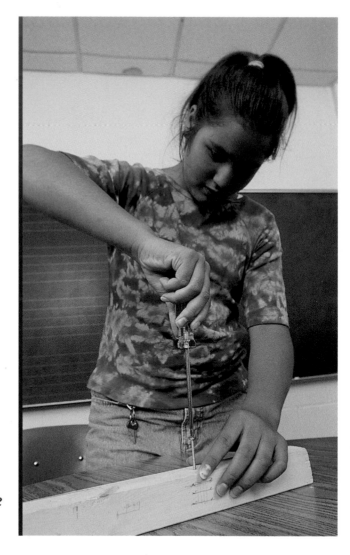

When you turn a screwdriver, you make its axle turn, too.

🌀 *Slowly roll the spool until the dot touches the paper again.*

Pull the pencil out of the spool. Draw a dot on the spool's edge. Then draw a dot on a sheet of paper. Line up the two dots. Slowly roll the spool until its dot touches the paper again. Mark the paper with another dot. The distance between the two dots on the paper is equal to one full turn of the spool. The wheel rolls a long distance in one turn.

Now make a dot on the side of your pencil.
Line it up with the first dot on the paper.
Slowly roll the pencil until its dot touches the
paper again. Mark the paper with another dot.

*Make a dot on the side of the pencil. You can use
chalk, a pen, or a marker.*

Starting
Point

🌀 *One turn of the pencil is a shorter distance than one turn of the spool.*

You can see that one turn of the pencil is a shorter distance than one turn of the spool. Remember, the pencil is the axle and the spool is the wheel. So one turn of the axle is shorter than one turn of the wheel.

Turning a large wheel is easier than turning a small axle. So your work is easier. You can prove this.

31

Try turning an axle instead of its wheel. It takes more force to do your work that way.

Get a screwdriver. Ask an adult to find a screw that you are allowed to turn. Hold the screwdriver's thin shaft. Try to unscrew the screw. It is hard to do. You must use a lot of force.

Try it again. But this time hold the handle of the screwdriver. It is probably easy to turn the screw. You turn the wheel a longer distance than you turned the shaft. But you don't have to use as much force. That makes your work easier.

It is easier to turn a wheel a long distance than it is to turn its axle a short distance.

Sometimes one wheel and its axle are all you need to do work. A unicycle has one wheel and axle. Riding a unicycle is fun. But balancing on it is hard. The rider can tip over easily.

Sometimes it helps to add more wheels and axles. Two wheels make balancing easier. But learning to ride a bike with two wheels is still hard. The bike tips over if you stop moving.

A tricycle is easy to ride. It has two back wheels. These wheels share one axle. A tricycle also has a front wheel and axle. These three wheels make a tricycle steady.

It's very easy to ride in a wagon. A wagon has two axles and four wheels. The four wheels make the wagon very steady. You have to work hard to tip over a wagon.

A tricycle has two back wheels and one front wheel. It is much easier to ride a tricycle than a unicycle.

35

Chapter 5

GEARS

Wheels and axles are many sizes and shapes. Some wheels are big. Others are small. Some axles are long. Others are short. Different wheels and axles are used for different jobs.

Some wheels have teeth. Teeth are bumps around the edge of a wheel. A wheel with teeth is called a gear.

Two gears work together. The teeth of one gear fit between the teeth of the other gear. When one gear turns, its teeth push against the teeth of the other gear. So the other gear turns too!

This is an eggbeater. The crank is attached to the big gear in the middle. A small gear is above each beater. The teeth of all three gears fit together. Turning the crank makes the big gear turn. This makes the small gears turn. Then the beaters spin fast.

What do you think will happen if someone turns the red gear?

Turning the red gear makes the blue gear turn, too. But the blue gear turns in the opposite direction from the way the red gear turns.

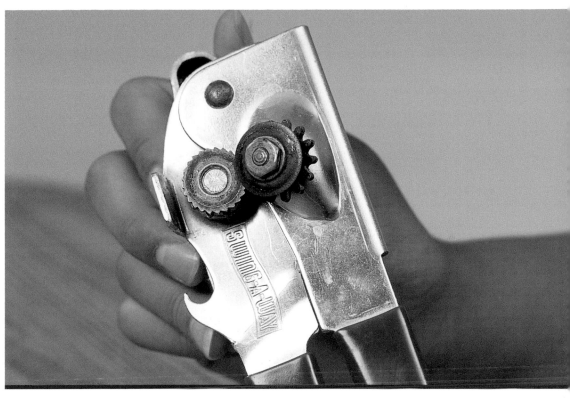

When you turn the handle of a can opener, the gears turn in opposite directions.

Look at a can opener. Squeeze the handles together. Notice how one gear's teeth fit into the spaces between the other gear's teeth. The knob is attached to one of the gears. Turn the knob. This action turns both gears. But the gears move in opposite directions.

Sometimes gears work together even though they do not touch each other. Look at the big gear near a bike's pedals. It is far away from the small gear in the back wheel. But a chain wraps around both gears. The teeth of the gears poke into the links of the chain. The chain connects the gears.

A bike has a large gear near the pedals and a small gear in the back wheel.

The pedals turn the big gear. The big gear turns the chain. The chain turns the small gear. Then the back wheel turns.

A bike's pedal is a big handle. The pedal turns the big gear. When the big gear turns, the chain moves with it. The links of the chain are hooked on the teeth of the small gear. When the chain moves, the small gear turns. When the small gear turns, the back wheel of the bike turns. Maybe it turns fast enough for you to win a race!

Now you know how the gears on these bikes work.

You have learned a lot about wheels and axles. Using a wheel gives you an advantage. An advantage is a better chance of finishing your work. Using a wheel and axle is almost like having a helper. That means you'll have time to do more work, like skating!

Wheels and axles make work easier!

ON SHARING A BOOK

When you share a book with a child, you show that reading is important. To get the most out of the experience, read in a comfortable, quiet place. Turn off the television and limit other distractions, such as telephone calls. Be prepared to start slowly. Take turns reading parts of this book. Stop occasionally and discuss what you're reading. Talk about the photographs. If the child begins to lose interest, stop reading. When you pick up the book again, revisit the parts you have already read.

Be a Vocabulary Detective
The word list on page 5 contains words that are important in understanding the topic of this book. Be word detectives and search for the words as you read the book together. Talk about what the words mean and how they are used in the sentence. Do any of these words have more than one meaning? You will find the words defined in a glossary on page 46.

What about Questions?
Use questions to make sure the child understands the information in this book. Here are some suggestions:

> What did this paragraph tell us? What does this picture show? What do you think we'll learn about next? What is force? How are simple machines different from complicated machines? What kind of force slows or stops moving objects? Can you find some machines in your home that have wheels and axles? How do gears work? What is your favorite part of the book? Why?

If the child has questions, don't hesitate to respond with questions of your own, such as: What do *you* think? Why? What is it that you don't know? If the child can't remember certain facts, turn to the index.

Introducing the Index
The index helps readers find information without searching through the whole book. Turn to the index on page 47. Choose an entry such as *gears* and ask the child to use the index to find out how gears work. Repeat with as many entries as you like. Ask the child to point out the differences between an index and a glossary. (The index helps readers find information, while the glossary tells readers what words mean.)

SIMPLE MACHINES

Books

Baker, Wendy, and Andrew Haslam. *Machines*. New York: Two-Can Publishing Ltd., 1993. This book offers many fun educational activities that explore simple machines.

Burnie, David. *Machines: How They Work*. New York: Dorling Kindersley, 1994. Beginning with descriptions of simple machines, Burnie goes on to explore complicated machines and how they work.

Hodge, Deborah. *Simple Machines*. Toronto: Kids Can Press Ltd.: 1998. This collection of experiments shows readers how to build their own simple machines using household items.

Van Cleave, Janice. *Janice Van Cleave's Machines: Mind-boggling Experiments You Can Turn into Science Fair Projects*. New York: John Wiley & Sons, Inc., 1993. Van Cleave encourages readers to use experiments to explore how simple machines make doing work easier.

Ward, Alan. *Machines at Work*. New York: Franklin Watts, 1993. This book describes simple machines and introduces the concept of complicated machines. Many helpful experiments are included.

Websites

Simple Machines
<http://sln.fi.edu/qa97/spotlight3/spotlight3.html> With brief information about all six simple machines, this site provides helpful links related to each and features experiments for some of them.

Simple Machines—Basic Quiz
<http://www.quia.com/tq/101964.html> This challenging interactive quiz allows budding physicists to test their knowledge of work and simple machines.

GLOSSARY

axle: a bar that goes through the center of a wheel

complicated machines: machines that have many moving parts

force: a push or a pull

friction: a force caused when two objects rub together

gear: a wheel with bumps around its edge

simple machines: machines that have few moving parts

wheel: a round object that turns on an axle

work: moving an object from one place to another

INDEX

axles, 22–28, 31

complicated
    machines, 10

force, 7–8, 14, 32–33
friction, 14–18, 20–21

gears, 36–41

simple machines, 11,
    22

wheels, 22–27, 31–37
work, 6–10, 31

About the Authors

Sally M. Walker is the author of many books for young readers. When she isn't busy writing and doing research for books, Ms. Walker works as a children's literature consultant. She has taught children's literature at Northern Illinois University and has given presentations at many reading conferences. She lives in Illinois with her husband and two children.

Roseann Feldmann earned her B.A. degree in biology, chemistry, and education at the College of St. Francis and her M.S. in education from Northern Illinois University. As an educator, she has been a classroom teacher, college instructor, curriculum author, and administrator. She currently lives on six tree-filled acres in Illinois with her husband and two children.

About the Photographer

Freelance photographer Andy King lives in St. Paul, Minnesota with his wife and daughter. Andy has done editorial photography, including several works for Lerner Publishing Group. Andy has also done commercial photography. In his free time, he plays basketball, rides his mountain bike, and takes pictures of his daughter.

METRIC CONVERSIONS

WHEN YOU KNOW:	MULTIPLY BY:	TO FIND:
miles	1.609	kilometers
feet	0.3048	meters
inches	2.54	centimeters
gallons	3.787	liters
tons	0.907	metric tons
pounds	0.454	kilograms